‹MOZAMBIQUE›

MAJOR WORLD NATIONS

MOZAMBIQUE

R. S. James

CHELSEA HOUSE PUBLISHERS
Philadelphia

Chelsea House Publishers

Contributing Author: Tom Purdom

Copyright © 1999 by Chelsea House Publishers,
a division of Main Line Book Co.
All rights reserved.
Printed and bound in the United States of America.

First Printing

1 3 5 7 9 8 6 4 2

Library of Congress Cataloging-in-Publication Data

James, R. S.
Mozambique.
Includes index.
Summary: Surveys the history, topography, people, and culture
of Mozambique, with an emphasis on its current economy,
industry, and place in the political world.
ISBN 0–7910–4744–X
1. Mozambique. [1. Mozambique.]
I. Title.
DT453.J36   1987
967'.9   87–11739
CIP

◄CONTENTS►

Map 6

Facts at a Glance 9

History at a Glance 11

Chapter 1 Mozambique and the World 15

Chapter 2 The Land 19

Chapter 3 Ancient Empires and Explorers 31

Chapter 4 Colonialism and Independence 43

Chapter 5 Country of the Good People 61

Chapter 6 Government and Social Services 75

Chapter 7 Economy and Transportation 81

Chapter 8 Cities and Towns 87

Chapter 9 Mozambique in Review 93

Glossary 97

Index 101

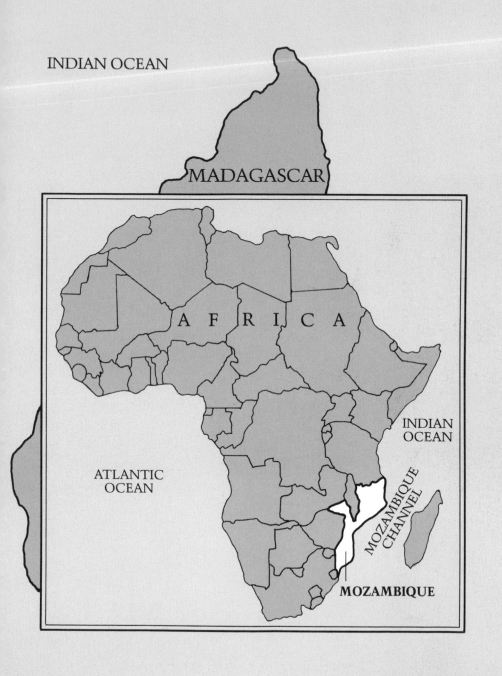

INDIAN OCEAN

MADAGASCAR

INDIAN
OCEAN

ATLANTIC
OCEAN

A F R I C A

MOZAMBIQUE CHANNEL

MOZAMBIQUE

◄ FACTS AT A GLANCE ►

Land and People

Area	309,600 square miles (801,590 square kilometers)
Highest Point	Mount Binga, 7,990 feet (2,436 meters)
Coastline	1,535 miles (2,470 km)
Greatest Length	1,100 miles (1,770 km)
Greatest Width	400 miles (640 km)
Major Rivers	Zambezi, Limpopo, Save, Changane, Buzi, Lurio
Capital	Maputo (population 1,700,000)
Other Major Cities	Beira (population 300,000+), Nampula (population 230,000)
Annual Rainfall	30 to 60 inches (750 to 1,500 millimeters)
Temperature Range	52° to 86° Fahrenheit (11° to 30° Centigrade)
Population	18,000,000
Population Density	57 people per square mile (22 per sq km)
Population Distribution	Rural, 87 percent; urban, 13 percent
Official Language	Portuguese
Literacy Rate	40 percent
Ethnic Groups	Indigenous groups (Makua-Lomwe, Tsonga, Makonde, others) amount to more than 96 percent of the population; the remainder include Europeans, Euro-Africans, and Indians
Religions	Indigenous beliefs, 50 percent; Christian, 30 percent; Muslim, 20 percent

Economy

Major Resources	Coal, titanium, natural gas, hydropower
Major Exports	Shrimp, cashews, cotton, sugar, copra
Major Imports	Food, clothing, farm equipment, petroleum
Agriculture	Employs 90 percent of work force, accounts for 33 percent of gross domestic product
Major Crops	Cotton, cashew nuts, sugarcane, tea, cassava, corn, rice, tropical fruits, beef, poultry
Currency	*Metical*, divided into 100 centavos

Government

Form of Government	Republic
Chief of State	President, elected by popular vote
Head of Government	Prime minister, appointed by the president
Legislature	Assembly of the Republic; members elected by popular vote for 5-year terms
Political Parties	Mozambique Liberation Front (Frelimo); Mozambican National Resistance (Renamo); Democratic Union (DU)
Administrative Organization	Ten provinces: Cabo Delgado, Gaza, Inhambane, Manica, Maputo, Nampula, Niassa, Sofala, Tete, Zambezia

◄HISTORY AT A GLANCE►

by 15,000 B.C.	Wandering groups of primitive hunter-gatherers form organized tribes.
1 to 300 A.D.	Bantu tribes migrate into southern Africa, bringing agricultural, herding, and ironworking skills.
after 300	Bantu peoples settle in the highlands and river valleys of Mozambique. They absorb the original inhabitants and create several large city-states.
700s	Arab traders begin to visit the Mozambique coast.
by 1000	Arab trading posts, which are linked to Arab states in the north, are established all along the coast.
1498	Portuguese navigator Vasco da Gama visits the coast on his way to India. He meets the Arab sheik of Mozambique Island.
1500	Another Portuguese explorer, Pedro Álvares Cabral, maps the Mozambique coast.
about 1505	Portugal establishes a settlement at the site of present-day Beira. The kingdom of the Mutapa people, in the interior of Mozambique, reaches its height.
1507	The Portuguese conquer Mozambique Island.
by 1514	The convict-explorer António Fernandes has mapped the region's interior.
1530s to 1540s	The Portuguese build a series of forts along the rivers leading into the interior. They hope to find gold in the fabled Mutapa kingdom.
1561	Gonçalo da Silveira, a Jesuit priest, reaches the court of the Mutapa king, where he is killed.

1569 to 1575	The Portuguese lead an unsuccessful military expedition against the Mutapa.
1580	The Zimba people attack the Mutapa from the north.
early 1600s	The Dutch make three unsuccessful attempts to seize Mozambique Island. The Rozwi kingdom from the south and the Maravi from the north attack the Mutapa and the Portuguese. After 1620, the Maravi hold off Portuguese advances for nearly a century.
1629	Mavura, placed upon the Mutapa throne by the Portuguese, signs a treaty giving his kingdom to Portugal.
1600s to 1800s	*Prazeros* (landholders of huge estates) rule much of Mozambique.
1700 to 1850	Peak years of the slave trade. Arab, Swahili, and Yao traders lead raids on many tribes.
1752	Portugal appoints the first governor of Mozambique.
1777 to 1782	Austria captures Lourenço Marques and rules the city temporarily.
1790s	France unsuccessfully tries to add Mozambique Island to its Indian Ocean territories.
1853 to 1856	Missionary-explorer David Livingstone traces the course of the Zambezi River from the interior to the coast.
1856 to 1864	Livingstone unsuccessfully tries to take a steamboat up the Zambezi River. He discovers Lake Malawi.
1880	The Portuguese government revokes all the prazeros' land claims and must fight for 15 years to subdue the powerful land barons.
1888	Lourenço Marques is made the colony's capital.
1889	Sir Harry Johnstone is appointed British consul to Mozambique and Malawi. He tries to abolish the last remnants of the slave trade.
1894 to 1912	Portugal sends military expeditions to subdue native tribes.

1917	The Makonde Rebellion is the last native uprising against Portuguese rule.
1930	The Colonial Act transfers governmental power from local officials to Portugal.
1951	Mozambique is made an overseas province, an official part of Portugal.
1950s	Independence movements arise and gain strength.
1960	Government troops kill 500 Makonde at a protest against forced labor.
1962	Eduardo C. Mondlane forms the Mozambique Liberation Front (Frelimo) to overthrow Portuguese rule.
1964	Fighting begins between Frelimo and Portuguese troops.
1969	Mondlane is assassinated.
1975	Frelimo seizes control of the government and declares Mozambique independent. Samora Machel is named president.
1976	Civil war begins between Frelimo and the forces of the Mozambican National Resistance (Renamo).
1986	President Machel dies in a plane crash. Joaquim Chissano becomes president.
1992	Cease-fire agreement ends 16-year civil war.
1994	Free election held. Head of Frelimo, Joaquim Chissano, elected president. Renamo participates in election, wins 112 seats in legislature.
1995	Repatriation of civil war refugees officially completed. Mozambique joins British Commonwealth.

Early explorers, impressed by the generosity of the land's friendly inhabitants, called Mozambique "The Country of the Good People."

Mozambique and the World

In 1498, a bold Portuguese navigator named Vasco da Gama set out in search of a sea route from Europe to India. Fighting fierce storms, he sailed around the southern tip of Africa and discovered unknown lands bordering the Indian Ocean. But whenever da Gama and his weary, starving crew went ashore to find food and water, they were attacked by hostile, frightened Africans. In desperation, da Gama sailed into the Zandamela region of what is now Mozambique (pronounced mo-zam-BEEK). One of his men wrote of the Mozambicans: "The chief said that we were welcome to anything in his country of which we stood in need. We called the country *Terra da Boa Gente*—Country of the Good People." This was the beginning of a 500-year-long relationship between Portugal and Mozambique.

After spending a month in Mozambique, da Gama sailed on toward his goal, the western coast of India. This historic voyage opened the great era of trade between Europe and the East. Other Portuguese explorers and adventurers followed him to the "Country of the Good People." Eventually, Portugal claimed ownership of Mozambique and ruled the country until the Mozambican people won their fight for independence in 1975.

At the head of the independence movement was Samora Machel, leader of the revolutionary Mozambique Liberation Front (Frelimo). After

Mozambique became independent, Frelimo set up a Socialist government (one guided by the belief that property should be owned by the state or other groups). Frelimo made Machel president of Mozambique, and he held that post until October 1986, when he was killed in a plane crash. Joaquim Chissano then became president.

As a Socialist nation inspired by Communist principles, Mozambique had close ties with the now-disbanded Soviet Union and with other Communist nations. It received substantial aid from the Soviet Union and had many technical and agricultural advisors from that country. Communist Cuba, too, did much to help the Mozambicans improve their production of sugarcane. Despite this assistance, Mozambique never became part of the Soviet-dominated Communist-bloc nations.

During his time in office, President Machel also welcomed advisors and aid from Western nations, such as Sweden and the United States. Mozambique became a member of the United Nations, the International Monetary Fund, and the Organization of African States. But relations between Mozambique and the United States have been troubled.

The first U.S. ambassador arrived in the new nation in 1976. During that year, the United States gave Mozambique $10 million in financial aid. One year later, however, U.S. officials feared that Mozambican police and military troops were violating human rights in their attempts to stop an armed anti-Frelimo group called the Mozambican National Resistance (Renamo). The United States said that it would no longer give aid to Mozambique unless Frelimo promised that human rights would be respected. After several years of arguing, Mozambique expelled four members of the U.S. embassy staff. The United States offered to appoint a new ambassador and to continue certain aid programs.

Communication between the two nations improved, and in 1983 a new United States ambassador arrived in Mozambique. That same year, Mozambique sent its first ambassador to the United States. In turn, the United States increased its financial aid. When Mozambique joined the World Bank and International Monetary Fund in 1984, Western aid soon

Government soldiers inspect the remains of a sawmill destroyed during the 16-year civil war.

became more important than Soviet assistance. In the late 1980s and early 1990s, as Communist states crumbled in Eastern Europe and the Soviet Union unraveled, Mozambique's government formally abandoned its adherence to socialism.

Like many other African nations, Mozambique has faced serious difficulties while trying to change from a colony into a modern nation. The Mozambicans have struggled against poverty, disease, drought, and famine. Their worst problem was the devastating 16-year civil war between Frelimo and Renamo. A cease-fire agreement in 1992 opened a new era, and in 1994 Mozambique held its first free election.

Today, the government of Mozambique is encouraging foreign investment and development of its mineral resources in order to build a solid economic base for growth. In the future, that growth should improve the lives of Mozambicans through better food production, health care, and education. If the government can reduce its dependence on foreign aid, it may have a chance to turn Mozambique into a thriving nation.

Mozambique's terrain includes both low, forested mountains and lush farmlands, such as this tea plantation near the western border of the country.

The Land

Mozambique is located on the southeastern coast of Africa. It is bordered on the north by Tanzania and on the south by Swaziland and the Natal region of South Africa. Its western neighbors are Malawi, Zambia, Zimbabwe, and the Transvaal region of South Africa. To the east lies the Mozambique Channel, which separates Mozambique from the island nation of Madagascar and the Indian Ocean.

With an area of 309,600 square miles (801,590 square kilometers), Mozambique is about twice the size of the state of California. The country is irregular in shape and about 1,100 miles (1,770 kilometers) long from north to south. It is approximately 400 miles (640 km) wide in the north and about half as wide in the south.

Mozambique encompasses many types of terrain: rain forests, swamps, sand dunes, mountains, grasslands, and dazzling white beaches with palm trees and coral reefs. Its two most prominent geographical features are Lake Malawi and the Zambezi River. Along with the country of Malawi, Lake Malawi stretches southward into Mozambique from the northwest, nearly dividing the country in two. The Zambezi River flows from west to east across the middle of the country and empties into the Mozambique Channel. The regions north and south of the river have somewhat different histories, climates, and cultures.

Almost half of Mozambique consists of a low-lying coastal plain that rarely rises to more than 600 feet (200 meters) above sea level. In the south, this flat, gently rolling plain extends across almost the entire width of the country, except where it is broken by the range of hills called the Serra da Gorongosa, located in the center of the narrowest part of the country, in the south central region.

Near the Zimbabwe border, the ground begins to rise toward a range of low mountains on the other side of the boundary. One high peak, however, is located just inside Mozambique. Called Mount Binga, it is 7,990 feet (2,436 meters) tall and is the highest point in the country. Not far from Mount Binga, also near the Zimbabwean border, is a large lake called Lake Oliveira Salazar. Many water birds live on its wide, marshy shores. Far to the south, along the borders with Swaziland and South Africa, the coastal plain ends in a chain of steep, rocky hills. They are called the Lebombo Mountains, although they are only about 1,500 feet (500 m) high—which is tall in comparison with the flat, swampy coastline of the region.

In the north, the coastal plain is narrower. Inland, plateaus or highlands rise from 500 to 2,000 feet (166 to 666 meters) above sea level. The terrain north of the Zambezi River is higher and more rugged than that of the south. It has many hills, rocky bluffs, steep streams, and dense forests, especially near Lake Malawi. Near the border with Malawi, the highlands rise into low, tree-covered mountains, with peaks ranging from 6,300 feet (1,913 m) to 7,900 feet (2,414 m) in height.

Lake Malawi, the third largest lake in Africa, is long and narrow in shape and covers 11,430 square miles (29,718 square kilometers), mainly within the country of Malawi. It lies in a huge trench called the Southern Rift, which is part of a great system of rifts, or valleys, that runs from the Middle East through eastern Africa. The rugged, rocky valley wall of the lake's southeastern shore is inside Mozambique's borders. This region is sparsely settled, although some farmers and fishermen live in small communities between the cliffs and the lakeshore.

Some 50 rivers flow eastward through Mozambique to empty into the Mozambique Channel. Chief among them are the Zambezi, the Limpopo, the Changane, the Save, the Buzi, the Lurio, and the Lugenda. The Shire River, Lake Malawi's only outlet, flows through a short stretch of Mozambique before joining the Zambezi. The Rovuma River makes up the boundary with Tanzania on the north. The largest rivers—the Zambezi, the Limpopo, and the Save—all originate to the west of Mozambican territory. The rivers that start in Mozambique are smaller, and some of them are seasonal, flowing only during the rainy season.

The Zambezi, 1,650 miles (2,640 kilometers) long, is Africa's fourth largest river. It begins in Angola, in the heart of the continent, and flows along the border between Zambia and Zimbabwe before entering Mozambique. In Tete, the westernmost province of Mozambique, the Zambezi

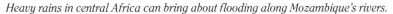

Heavy rains in central Africa can bring about flooding along Mozambique's rivers.

once thundered over the steep Cabora Bassa Falls, but a huge dam and power plant were built there between 1969 and 1976. The dam caused the Zambezi to back up into a lake 150 miles (240 km) long and 600 square miles (1,560 square kilometers) in area. The lake devoured many villages, and more than 25,000 people were forced to move to new homes. Today, Cabora Bassa is one of the largest hydroelectric plants in Africa, and the lake has been stocked with fish to improve the diet of the people who live in this poor and remote part of the country.

The Zambezi is navigable for 280 miles (448 kilometers), from Cabora Bassa to the Mozambique Channel. Wood-burning steamboats, motor launches, and canoes carry passengers and cargo back and forth along the lower reaches of the river. The delta, where the river meets the sea, is a treacherous maze of shifting sandbanks and confusing channels. An unusual feature of the Zambezi Delta is the existence of the *baixos*, or floating islands—masses of vegetation washed downstream by heavy rains. The baixos sometimes drift around the river mouths for many years and are additional hazards to navigation.

Although Mozambique is only about 1,100 miles (1,770 kilometers) long from north to south, its coastline actually runs for 1,535 miles (2,470 km) because it contains so many bays, coves, and capes. At its southern tip is Delagoa Bay, a large harbor that is one of the finest natural harbors in the world and is the site of Maputo, the nation's capital. From Delagoa Bay north to the Zambezi, the coast is bordered by dunes and by swamps of low, twisted mangrove trees, which grow in dense clusters along beaches and river mouths. North of the Zambezi Delta, the coast is less swampy, consisting of sandy beaches and rocky cliffs.

Many small, coral islands lie just off the shore of northern Mozambique, and coral reefs line the coast. Midway between the Zambezi Delta and the northern border of the country is Mozambique Island, the ancient settlement, visited by da Gama, that gave the country its name. The chain of islands continues to Cabo Delgado, or Delgado Cape, a high, rocky promontory that is the northernmost point in Mozambique.

Tiny coral islands off the coast are uninhabited, but fishermen sometimes land on them.

The east coast of Mozambique and the west coast of the nearby island of Madagascar look as if they would fit together like puzzle pieces, because Madagascar was once part of Africa, joined to Mozambique. Hundreds of millions of years ago, the same forces that created the rift valleys pulled Madagascar slowly eastward. Scientists believe that all of the earth's continents rest on rock plates that float on a sea of molten lava deep beneath the planet's surface. This theory, called plate tectonics, suggests that the continents shift into new positions over very long periods of time. The similarity between the coasts of Mozambique and Madagascar led scientists to examine the rocks and fossils of both countries, and the geological evidence they found to connect the two places helped them to form the plate tectonics theory.

The Mozambique Channel, which lies between Mozambique and Madagascar, varies in width from 250 miles to 600 miles (400 to 960 kilometers). The Mozambique Current, a stream of warm water from the equatorial regions of the Indian Ocean, flows south through the channel. Its warm water supports tropical ocean life, including coral reefs, warmwater fish, porpoises, and sharks. A large number of giant manta rays inhabit the waters of the Zambezi Delta. These rays sometimes upset canoes and even good-sized fishing boats.

Climate

Mozambique has a somewhat dry, tropical climate. Temperatures are warm year round, but rainfall is lighter than in many other tropical areas.

Because Mozambique lies south of the equator, its summer and winter climates are the opposite of those of the United States and Europe. Summer, the rainy season, lasts from November through April. Temperatures at this time of year range from 79° to 86° Fahrenheit (26° to 30° Centigrade) in the lowlands and from 71° to 77° F (22° to 25° C) in the highlands. The dry season, which lasts from May until October, is slightly cooler, with temperatures ranging from 59° to 68° F (15° to 20° C) in the lowlands and from 52° to 59° F (11° to 15° C) in the highlands.

Average annual rainfall ranges from 30 inches (750 millimeters) in the capital city of Maputo to about twice that at Beira. In some years, rainfall is even lighter. Droughts are common, especially in the south central part of the country. The sandy soil of this region causes rainwater to evaporate or drain away easily. On the other hand, floods sometimes occur in the lowlying river valleys when heavy rainfall in the interior of the continent causes the major rivers to swell. North central and northwestern Mozambique generally have the most consistent rainfall. These steep and hilly regions rarely experience either water shortages or floods.

Like many Indian Ocean countries, Mozambique is subject to occasional violent tropical cyclones. These hurricane-like storms begin far out at sea and fall upon the coast with devastating power, bringing gale winds, high waters, and sometimes immense tidal waves. Storms on the Mozambique coast often feature as many as a dozen waterspouts, twisting skyward from the surface of the sea like huge, gray snakes. In years past, these storms killed thousands of people and caused great damage to houses, crops, and fishing fleets. Mozambicans called them "The Whips of God." Today, however, meteorologists use satellite weather stations to watch for the first signs of a cyclone. Their storm warnings help save Mozambican lives and property.

Plants and Animals

Mozambique has several types of vegetation. In the narrow ravines of the northern mountains and along the upper reaches of the river, large trees rise a hundred feet or more above the rain forest floor. Among these trees are the Senegal *khaya*, a large, dark hardwood similar to mahogany, and the *mopani*, a form of ironwood that produces durable timber. The trees trail vines and orchids and create a dark green gloom under the canopy of their leaves.

Much of southern Mozambique is savanna country (sometimes called "veld" or "veldt"): open, rolling grassland with occasional patches of brushy forest. Ironwood and ebony trees dot the savanna slopes, and the most common tree is the baobab. The silhouette of the baobab, with its thick trunk and low, twisting, wide-spreading branches, is one of the most common sights in southeastern Africa.

The baobab tree's thick trunk and twisting branches form a common African silhouette.

Along the coast, coconut palms are abundant, especially near the Zambezi Delta. Date palms also grow in the coastal region and on the coral islets. Almond trees with reddish brown leaves, bamboo, papyrus (the reed used by the ancient Egyptians to make paper), and spear grass grow in the wetter areas near the river. Water hyacinths, which sometimes form great, island-like clumps in the water, are also found in the region. A tall variety of screw pine that is abundant around the Zambezi Delta is named *Pandanus livingstonianus,* in honor of the Scottish missionary-explorer David Livingstone, who traveled up the Zambezi in the 1850s. The offshore islands have a few palms and some low casuarina trees (a tropical pine). Many trees on the islands are tilted or twisted westward by the Indian Ocean trade winds that blow from the southeast.

Mozambique is rich in wildlife. The savanna region around the Save River is the home of big game, including lions, elephants, giraffes, and black rhinoceroses. White rhinoceroses are sometimes spotted in Mozambique, but they are becoming quite rare. Herds of zebra and buffalo roam the plains, and there are many types of antelope: elands, impalas, nyalas, oribis, hartebeests, kudus, and tiny duikers. Monkeys and warthogs (wild pigs with sharp, curved tusks) live in the forests of the savanna. Jackals, wild dogs, spotted hyenas, and mongooses (small, snake-eating rodents) all inhabit central Mozambique, as do cheetahs, servals (spotted wildcats about the size of small dogs), and civets (cats often hunted for their scent glands, which are used in some perfumes and native medicines).

The lower reaches of the country's rivers are full of dark gray or brown lumps that look like floating logs. The large ones are hippopotamuses. They submerge themselves on lake or river bottoms, sometimes for as long as an hour, and scoop water plants into their huge jaws. The smaller lumps are crocodiles. Hippos can be vicious if they are startled or injured, but they generally leave humans alone. Crocodiles, however, are far more dangerous. Ranging in length from 6 to 15 feet (2 to 5 meters) and equipped with snapping jaws and razor-sharp teeth, crocodiles kill or injure hundreds of Mozambicans each year.

A lump that looks like nothing more than a log floating in the river may really be a large-mouthed hippopotamus.

Other reptiles are also abundant. Poisonous cobras, puff adders, and vipers are found in Mozambique. The largest snake is the giant African python, which hangs from tree branches or lurks in abandoned wells or buildings. It can be as long as 12 feet (4 meters) and as thick as a man's arm, and it kills its prey—usually birds and small mammals—by wrapping its coils around it and squeezing. Pythons seldom attack humans.

Huge flocks of flamingos inhabit the northeastern part of Mozambique, in the area known as Nampula. Other water birds, such as ibises, herons, cranes, pelicans, ducks, and Egyptian geese, live on the shores of Lake Malawi and along the country's waterways. Eagles, crows, hawks, buzzards, and vultures soar above the savannas. Smaller birds, including guinea fowl, quail, and partridges, nest in the thick brush at the edges of forest regions.

Lake Malawi and the rivers of Mozambique contain many kinds of fish, some of which are edible. Fishermen especially prize the *tilapia*, a perchlike fish with flavorful, white meat. Catfish are common in the rivers.

They sometimes grow to enormous sizes; the Zambezi and Limpopo Rivers have yielded catfish 6 feet (2 meters) long. The coastal waters contain crabs, lobsters, shellfish of various sorts, tuna, marlin, swordfish, garfish, grouper, barracuda, squid, and sometimes schools of flying fish. These fish don't actually fly, but they do look like silver birds when they leap out of the water for great distances between the crests of waves. Smaller fish—wrasses, angelfish, parrotfish, and others—live among the coral reefs in huge, brilliantly colored schools.

Mozambique is one of the world's finest sources of seashells. Specimens of almost every seashell species found in the Indian Ocean and the western Pacific Ocean wash up on Mozambique's beaches. Some of the shells are extremely rare. One example is *Cymatium ranzanii*, a knobby, yellow, orange, and brown shell about 8 inches (20 centimeters) long. Of the ten known specimens of *Cymatium ranzanii* in the world's museums, five are from Mozambique.

Like most African wildlife, the animals of Mozambique are threatened by the area's rapidly growing human population. Men hunt wild animals for food and also for valuable substances, such as elephant ivory, crocodile skin, and rhinoceros horn (which is ground into a powder and sold as a form of folk medicine in the East). Even animals that are protected by law, such as the rhinoceros, are not safe from poachers with high-powered rifles. In addition, the spread of cities and farmlands means that the forests and open plains of Africa are shrinking, giving the animals less and less room to roam.

But unlike the wildlife of neighboring countries, Mozambique's wildlife has not suffered from prolonged drought. And the Mozambican government has taken steps to preserve the country's animals by creating several national parks and wildlife preserves. The largest is Gorongosa National Park, in the Serra da Gorongosa. It consists of 2,190 square miles (5,690 square kilometers) of varied terrain, including hills, rain forest, savanna, and swamp. Although many national parks in Africa are larger, Gorongosa is one of the continent's most complete and well-

The government has taken steps to stop the slaughter of elephants for their ivory.

managed parks. Because many African animals travel in migratory herds and cover great distances, Gorongosa's wildlife sometimes wanders into Kruger National Park in South Africa, and animals from Kruger show up in Gorongosa. The two nations have agreed to protect each other's wildlife.

The Mozambican government plans to establish more wildlife preserves and parks when funds are available, including a park to preserve stretches of the offshore coral reef and its colorful undersea life. It also hopes to train and equip more conservation officers to study and protect wildlife. Mozambique strictly controls big-game hunting and is cooperating with neighboring countries to limit the smuggling of illegal skins and tusks across national borders. Some species are now thriving; black rhinos, which are becoming rare in many African countries, are increasing in Mozambique.

VASCO DA GAMA

Portuguese navigator Vasco da Gama was the first European to reach Mozambique. After battling fierce storms, he and his crew landed on the East African coast.

Ancient Empires and Explorers

People have lived in Mozambique since almost the dawn of the human race. Anthropologists believe that mankind originated in East Africa, in the regions that are now Kenya and Ethiopia, hundreds of thousands of years ago. Some groups of people migrated southward. These people became the first Mozambicans.

Mozambique's early inhabitants were small men and women who lived in caves. They were hunter-gatherers who did not practice agriculture or herding, but survived by eating wild fruit, berries, roots, and small animals. They left behind hundreds of fossil bones and stone axe-heads and knives throughout western Mozambique, Zimbabwe, and South Africa.

Between 15,000 and 40,000 years ago, these wandering groups of people gathered into larger, more organized tribes whose physical appearance and culture were much like those of the Pygmy people of present-day Africa. Like the Pygmies, they used bows and arrows, often tipped with poison, to kill antelope and other prey. They also built traps and snares for game and used hooks, spears, and wicker baskets to catch fish.

Over thousands of years, these people created hundreds of magnificent paintings in caves and on rock walls using vegetable dyes and charcoal. These paintings reveal that the men hunted in groups while the women gathered grain and fruit, and that the tribes migrated from place to place, following the vast animal herds that roamed the plains of Stone Age Africa. The very last paintings, 2,000 years old, depict iron tools and herds of sheep—hinting at an event that was about to transform human society in southeastern Africa.

That event was the arrival of the Bantu-speaking peoples. The Bantu were a cluster of related tribes from north central Africa. They were taller and darker-skinned than the original inhabitants of the south. For some reason unknown to historians, the Bantu began to multiply rapidly in the 1st century B.C. Large numbers of them migrated southward and settled most of southern Africa between the 1st and 4th centuries A.D. The Bantu were herdsmen and farmers who brought domesticated animals and agriculture into Mozambique. They also knew how to forge iron, from which their warriors fashioned iron spears. Their arrival in Mozambique marked the end of the country's Stone Age and the beginning of the Iron Age.

The Bantu newcomers outnumbered the original inhabitants and gradually absorbed them into their culture through conquest and inter-marriage. Today, all of Mozambique's native Africans belong to Bantu tribes. From the 4th century on, the Bantu settled in the highlands and

The Bantu ancestors of these children migrated into southeast Africa from the north.

river valleys of western Mozambique. There they herded cattle and sheep, mined gold and tin, hunted elephants for ivory, farmed millet and sorghum (two grains native to Africa), and made pottery. They built stone cities in the hills, where the kings and their wives lived. Workers and slaves lived in grass huts outside the city walls.

The greatest Bantu empire of Mozambique was the Mutapa (or Matapa) empire. Its heartland was along the Zambezi River, south of Lake Malawi. At its peak in the early 16th century, the Mutapa empire reached into what is now Zimbabwe and consisted of many cities and forts scattered over a wide region of forested hills. The capital city housed several thousand people. The emperor, called the mwene Mutapa, controlled several gold mines at secret locations.

The Mutapa and other kingdoms and tribes of the interior grew rich through trade with Arab merchants. Arabs had begun to visit the coast of southeast Africa as early as the 8th century A.D., traveling the Indian Ocean in swift cargo boats called dhows. By the 11th century, Arab trading posts and settlements lined the coast as far south as the Zambezi, and Arab city-states had risen on the coast north of Cape Delgado. Through the Arabs, African chieftains in the interior obtained goods from the Indian Ocean trade, such as porcelain vases and bowls from China, rice and spices from Indonesia, and silks from India. In turn, Arabia and the East received gold, ivory, and slaves from Africa.

Between the 11th and the 15th centuries, the Arabs strengthened their control of the coast. The Swahili, a people of mixed African and Arab heritage, introduced the Islamic religion and many Arabic words into Mozambican culture. By the late 15th century, Arab chieftains called sheiks ruled Mozambique Island and many of the other offshore islands. Tended by Swahili servants and African slaves, the sheiks lived in spacious houses built of coral blocks and surrounded by groves of orange and lemon trees imported from Arabia.

Vasco da Gama and his crew were the first Europeans to visit Mozambique. When they arrived in 1498, they landed in the Zandamela region,

Arab slave dealers used small sailing boats called dhows to carry their human cargo.

near the Limpopo River (which da Gama named the River of Cobras after the many snakes he saw there). Next they landed near the Zambezi Delta, and finally they moved north to Mozambique Island, where da Gama spent a month as the sheik's guest. The two quarreled because da Gama refused to give his ship's cannon to the Arabs. When he left the island bound for India, however, the sheik lent him an experienced Swahili pilot.

Da Gama noted that 15th-century Mozambique society was divided into two parts: the black African tribes of the interior and the southern coastal regions, and the Arab and Swahili traders along the central and northern coasts. The wealth of the coastal Arabs convinced the Portuguese that deep within Mozambique lay great treasures. In reality, the gold mines of the western hills were almost empty by the time of da Gama's arrival. But rumors of wealthy empires and cities of gold hidden deep in the rain forests lured many Portuguese explorers and adventurers to Mozambique.

In July 1500, a Portuguese navigator named Pedro Álvares Cabral stopped in Mozambique on his way to India. He drew a map of the coastline that confused many later explorers because it was so inaccurate. (Some historians believe that Cabral may have drawn the map badly on purpose, to keep others from making safe landings on the coast.) Either Cabral or another explorer, João da Nova, visited Delagoa Bay between 1500 and 1502 and left behind some stone inscriptions. In about 1505, the Portuguese established a settlement called Sofala near the present-day site of Beira. In 1507, they attacked and conquered Mozambique Island. The era of Portuguese domination had begun.

More than anything else, the Portuguese wanted to reach the Mutapa kingdom and take over its gold mines, but their first efforts were unsuccessful. Disease and hostile Africans killed many fortune hunters. Finally, the explorers sent a convict named António Fernandes into the rain forest with orders to find Mutapa. Although he never reached Mutapa, Fernandes succeeded in mapping much of the Mozambican interior by 1514.

Meanwhile, Arab and Swahili merchants continued to trade with the Mutapa by using the Zambezi River route. During the 1530s, the Portuguese established forts on the Zambezi at Sena, 160 miles (256 kilometers) upstream, and at Tete, 260 miles (416 km) upstream. In 1544, they built a fort at Quelimane to control traffic on the Zambezi Delta. At last, in 1561, a Portuguese Jesuit priest named Gonçalo da Silveira succeeded in reaching the court of the mwene Mutapa, whom he immediately baptized as a Christian. But Arab merchants at the court persuaded the mwene to strangle Silveira. In return, the Portuguese declared war on the Africans.

In 1569, a military expedition left Lisbon, Portugal, intent on seizing the Mutapa mines. But the expedition ended in 1575, after many of the

Portuguese soldiers died of disease and wounds from poisoned arrows. In the meantime, the mwene Mutapa faced another enemy. A warlike tribe called the Zimba had begun attacking the Mutapa from north of the Zambezi. After 1580, the Zimba assaults grew fiercer; the invaders also attacked Portuguese settlements along the Zambezi and on the coast. Unable to repel the Zimba, the mwene was forced to ask the Portuguese for help in 1601.

In return for their military aid against the Zimba, the Portuguese forced the mwene to sign a treaty giving Portugal rule over the territory. They then deposed the mwene and, in 1629, installed a new ruler named Mavura. Mavura supported the Portuguese claim to sovereignty over the Mutapa.

For the next several centuries, the Portuguese tried to establish control over Mozambique's interior while the African tribes fought among themselves and against the Europeans. In the 17th century, a group called

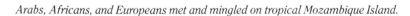

Arabs, Africans, and Europeans met and mingled on tropical Mozambique Island.

the Changamire formed a kingdom called Rozwi, south of the Zambezi. Hoping to take over the Mutapa throne, the Rozwi swept down the river. They forced the Portuguese to retreat and reduced the Mutapa empire to a small chiefdom in the lower Zambezi Valley. The last remnants of the Mutapa empire disappeared in the late 19th century.

While the Rozwi were attacking south of the Zambezi, a people called the Maravi attacked the Mutapa and the Portuguese from north of the river. In about 1620, a powerful Maravi chief united the entire north bank under one leader. For almost a century, the Maravi blocked European attempts to explore the interior. In the early 1700s, however, the confederation of Maravi tribes broke up. There was no longer a single large African power to halt the Portuguese advance.

By this time, however, the Portuguese had discovered how inhospitable the African interior could be. Instead of complete military control, they now wanted only the economic advantages of trade. The Portuguese set up outposts and annual fairs throughout the country and traded Portuguese cloth, glass beads, and firearms for the treasures of the African chieftains.

Adventurous settlers and traders from Portugal and Goa (a Portuguese colony on the coast of India) bought grants to huge tracts of land from local rulers. These private estates were called *prazos*, and the landholders were called *prazeros*. Eventually, the king of Portugal confirmed the ownership rights of the prazeros. In effect, the prazeros ruled much of inland Mozambique from the 17th through the 19th centuries. They maintained private armies of African and Portuguese soldiers, called *chikunda*. They interfered in local African politics and warred among themselves. The prazeros also refused to acknowledge the authority of the Portuguese government. They were like medieval land barons, each with his own small kingdom.

By the beginning of the 18th century, the supply of gold from the fabled Mutapa mines had become a trickle, the ivory trade had increased, and the Portuguese had taken over and expanded the slave trade that the

Prazeros *kept African families, servants, and private armies on their vast estates.*

Arabs had carried on for centuries. They captured or purchased men and women from chieftains in the interior and marched them to the slave markets at Pemba and Quelimane. The Yao, a tribal group living east of Lake Malawi, obtained guns from the Arabs and Portuguese and became the leading slave raiders of northern Mozambique.

No one knows how many slaves the Portuguese took from Mozambique between 1700 and 1850, when the trade was at its height. The traders sold some of the slaves to Brazil, Portugal's colony in South America, but shipped most of them to French and Portuguese plantations on the Indian Ocean islands of Réunion, Mauritius, Madagascar, and Zanzibar. By 1817, many countries north of the equator had abolished slave trading, but at the same time, the demand for slaves from southern ports increased. The chikunda soldiers of the Zambezi Valley warlords continued to deliver whole tribes to the slave dealers in the coastal cities. In the 1840s, the Dutch settlers in South Africa purchased many of these slaves.

Great Britain and other European nations officially abolished slave trading in southern Africa in 1836 and slavery itself in 1856. But these official proclamations had little real effect in Mozambique. Dealers no longer sold slaves outside of the country, but landowners within Mozambique continued to raid the tribes for unpaid labor to work on their plantations, transport ivory, and gather wild rubber. These practices continued until the end of the 19th century.

The Missionaries

The Portuguese introduced the Roman Catholic faith to the coastal areas, but made no widespread attempt to convert the Mozambicans to Christianity. This left the way open for Protestant missionaries, who were very active in Africa from 1800 on. The most vigorous missionaries were those sent out by the Church of Scotland, and the most famous of these was David Livingstone, who explored the Zambezi River and discovered Lake Malawi.

Livingstone began his missionary work in South Africa in 1841. He soon developed a taste for travel in unexplored districts. In 1851, in what is now Zambia, he came upon a large river not far from its source. It was the Zambezi, and in 1853, he set out to trace its entire course. His journey took him north into Angola and then south and east to Quelimane, which he reached in 1856. Upon returning to London after this stupendous achievement, Livingstone was hailed as a hero. He soon decided to return to the Zambezi, which he envisioned as a highway by which both Christianity and commerce could enter the heart of southeastern Africa. Portugal was not very interested in Mozambique at the time; internal strife and the Brazilian colony demanded more of the government's attention. As a

Livingstone's explorations inspired other Europeans to brave the African wilderness.

result, Livingstone had no trouble mounting a large expedition into Portuguese territory.

The expedition arrived at the Zambezi Delta in 1858, complete with a large paddle-wheel steamboat, the *Ma Robert*. For six years, Livingstone tried in vain to navigate the river. He encountered innumerable obstacles, including sandbanks, low water, disease, and hostile natives. Finally, he reached the Cabora Bassa Rapids, 30 miles (48 kilometers) of thundering cataracts that no boat has ever survived. Livingstone had expected to sail through the rapids but was forced to turn back in defeat. He then tried to take his boat up the Shire, the major tributary of the Zambezi, but he encountered rapids there, too. Exploring on foot, he discovered Lake Malawi and tried to set up a mission on its shores, but the local tribespeople, fearing that he was really a slave trader, drove him away. His expedition retreated from Mozambique in 1864, but Livingstone did not lose his taste for adventure.

Two years later, Livingstone set out in search of the source of the Nile River. He disappeared into central Africa and was believed dead until 1871, when explorer Henry Stanley found him there and greeted him with the famous phrase, "Dr. Livingstone, I presume?" In May 1873, Livingstone died on the shores of Lake Bangweolo in Zambia.

Livingstone's career inspired other missionaries to enter southeast Africa. In 1875, the Church of Scotland established the Livingstonia Mission near Lake Malawi. The missionaries detested the slave trade and hoped that Great Britain would take an active role in forcing Portugal to improve conditions in its colony. In 1883, they persuaded Great Britain to appoint a consul to Mozambique. A few years later, Great Britain engaged in battles with the Arab and Portuguese slavers in Mozambique and at other points on the African coast.

In 1898, Great Britain appointed Sir Harry Johnstone as consul to Mozambique and Malawi. Under his urging, the British became more involved in Mozambican affairs. They succeeded in stamping out the last traces of the slave trade near the end of the 19th century.

As they did during the colonial period, most Mozambicans today live by tilling the soil. The Portuguese found no gold mines or other riches in their colony.

Colonialism
and Independence

Other European nations did not accept Portugal's claim to Mozambique without question. They, too, were eager to obtain overseas colonies. In the early 1600s, the Dutch made three attempts to seize Mozambique Island but failed each time. The Austrians captured the port city of Lourenço Marques on Delagoa Bay in 1777, but Portugal recaptured it in 1782. During the 1790s, the French acquired many territories in the Indian Ocean and tried several times to take Lourenço Marques and Mozambique Island. During the same time, Mozambique's coastal islands suffered many cruel raids by Sakalava tribesmen from Madagascar.

Despite these attacks, Portugal managed to retain control of Mozambique. At first, it regarded the colony as part of the Portuguese Indian colony of Goa and allowed the governor of Goa to administer it. But in 1752, Mozambique received its own governor, who lived in Lourenço Marques. Portugal still did little to develop Mozambique, however, and sent few settlers there.

Near the end of the 19th century, the discovery of huge deposits of gold and diamonds in South Africa set off a series of territorial claims and counterclaims. Mozambique acquired new importance in the eyes of its Portuguese masters. In 1880, the government of Portugal declared that the prazeros' land grants had been revoked and that their territories belonged to the crown. The prazeros ignored the new law. It took four military expeditions, over the course of 15 years, to subdue the powerful landowners.

In 1884 and 1885, representatives of European nations met in Berlin to settle their disagreements over African frontiers. Some countries, including Portugal, claimed rights to areas that their countrymen had discovered. Other nations, such as Germany, felt that ownership should go to the country that occupied and used the region, not to the country that had discovered it. Portugal hoped to maintain its claim to all of south central Africa, which would link the west-coast colony of Angola to the east-coast colony of Mozambique. But as a result of the Berlin meeting, Great Britain acquired the lands that are now Zimbabwe and Malawi, thus breaking up the Portuguese holdings. In 1891, Great Britain and Portugal signed a treaty that limited Portugal's claims. Germany, too, acquired territory that Portugal had wanted, north of the Rovuma River.

The other European nations thought that Portugal was too weak and unstable to hold its African colonies. Great Britain and Germany even drew up a secret treaty dividing Portugal's holdings between them. But

In the 19th century, the nations of Europe snatched up colonies all over the world.

Portugal managed to retain control of Mozambique. In 1888, it designated Lourenço Marques the colony's capital, and during the 1890s it began to tighten its control over the interior districts. The Portuguese fought military campaigns against the Mozambicans almost every year from 1894 through the turn of the century, and the tribes in the northwest resisted them until 1912. The last protest against colonial rule was the Makonde Rebellion of 1917, led by the Makonde tribe, Mozambique's fiercest warriors, who inhabited an isolated plateau inland from Mozambique Island.

Once it had established control of the interior, the Portuguese government issued concessions—permits that allowed investment companies to operate enormous farms in Mozambique in return for rent. These companies set forced laborers to work planting and tending tea, sugar, cotton, coffee, and cacao. Some of these plantations became profitable. As a result, Portugal took back the concessions north of the Zambezi River in 1929 and those south of the river in 1942.

The Portuguese ran these plantations according to a system called *shibalo*, from the Swahili word *shiba*, meaning "serf." They forced the Africans to live and work on the plantations, forbade them to grow food crops or keep animals of their own, and permitted them to tend only cash crops, such as cotton and tea. For example, more than 500,000 workers in the northern regions were forced to grow cotton for export to Portugal instead of food for use at home. They had to buy provisions from company or government stores.

Although they were profitable, the plantations were not Mozambique's greatest source of income during the colonial era. The colony had two other valuable commodities: manpower and access to the sea. Manpower was the more useful, and the Portuguese government exploited it as much as possible. When slavery became illegal, Portugal began exporting hired laborers, mostly to the gold and diamond mines in the region of South Africa called Witwatersrand, or the Rand.

In a series of contracts called the Mozambique Conventions, begun in 1875 and renewed until 1964, Portugal and South Africa agreed that laborers from Mozambique would work in the mines. South Africa paid part of their wages directly to the Portuguese government in gold; Portugal then sold the gold on the international market and paid the Mozambicans in local currency after they returned home—sometimes years later. This method allowed Portugal to make a profit on the gold and gave South Africa an endless supply of labor. Because many Mozambicans were unable to make a living working on the plantations that had taken over their former tribal lands, they were willing to go to South Africa to earn even partial wages. By 1897, more than half of the 180,000 workers in the Rand came from Mozambique, most of them from the southern provinces. From 1901 on, between 100,000 and 150,000 Mozambicans went to work in South Africa each year.

In addition to manpower, Mozambique began to sell transportation services. The government built railways to carry freight between land-locked countries such as Rhodesia (now Zimbabwe) and the Mozambican seaports. Under the Mozambique Conventions, South Africa agreed to send part of its railway traffic through Mozambique.

World War I brought hard times to Mozambique. Tribesmen fought in military campaigns against the Germans in East Africa, and many died from war wounds or diseases caught from the white men. Famines occurred throughout the country. But after the war, Mozambique's increasing prosperity began to attract settlers from Portugal. The most ambitious settlement project was the Colonato Limpopo, a government-sponsored community of 10,000 Portuguese families on a 250,000-acre (10,000-hectare) site in the Limpopo Valley. Other Portuguese immigrants settled in the coastal cities and practiced trades, opened small businesses, or staffed the colonial administration.

In 1930, the colony had only about 17,000 whites; by 1950, there were 48,000 whites. By the mid-1970s, on the eve of independence, that number had increased to more than 250,000. For many years, the whites had little contact with the native inhabitants of the country, because the structure of colonial society kept the races separate. A law passed in 1927

The Portuguese used the lower reaches of the Zambezi and Limpopo rivers as highways.

divided all Mozambicans into two groups. One group, called *indigenas*, consisted of all but a few Africans. The indigenas were required to pay taxes, to serve on public and private labor gangs, to keep a curfew, to farm as ordered by the authorities, and to carry identification papers.

The other group, the *não indigenas*, consisted of Europeans, Asians, and a few Africans who were called *assimilados* (an African could become an assimilado if he spoke Portuguese, could give references to his good character, and gave up his traditional way of life for a job in business or industry). The não indigenas had all the rights of Portuguese citizens. Although the colony abandoned the indigena system in 1961 and granted all Mozambicans Portuguese citizenship, Africans continued to resent the injustices they had suffered.

In 1930, the Portuguese government passed the Colonial Act, which took some powers away from the administrators in Mozambique and tied the colony more closely to Portugal. In 1951, it changed Mozambique's status to that of an overseas province instead of a colony. Many natives, however, were not pleased with being joined to Portugal. They began to think about independence.

Frelimo

Independence movements swept through Africa in the years after World War II, and many colonies were granted independence—or seized it by force. Portugal refused to consider giving independence to Mozambique, but some Mozambicans were willing to fight for it. During the 1950s, more and more of them spoke up against Portuguese rule.

In 1960, a group of Makonde peasants staged a peaceful demonstration to protest being forced to grow cotton. Portuguese troops shot more than 500 of them. The survivors fled into Tanzania. There they joined a group of educated Mozambicans who had left their homeland because they opposed Portuguese rule. Their leader was Eduardo C. Mondlane, who had taught anthropology at Syracuse University in New York. In 1962, Mondlane formed the group into the Mozambique Liberation Front, or Frelimo. He and other Frelimo leaders pledged to drive the Portuguese out of Mozambique.

Fighting broke out between Frelimo forces and government troops on September 25, 1964. (Today, Mozambique celebrates that day as the beginning of its struggle for freedom.) One year later, Frelimo claimed to have an army of 8,000 in the northern provinces of Cabo Delgado and Niassa. In response, Portugal sent 70,000 Portuguese troops to smash the rebellion. Hoping to win international support for its action, Portugal opened up new investment opportunities in Mozambique, including the Cabora Bassa hydroelectric project and the British-owned Sena Sugar Estates. It also introduced some educational and economic benefits for native African Mozambicans. But these efforts were too little and too late. Mozambicans throughout the country supported Frelimo, even after Mondlane was assassinated in 1969.

Under the leadership of Samora Machel, one of Mondlane's followers, Frelimo carried the rebellion into Tete province in 1971. Portugal

(continued on page 57)

SCENES OF
MOZAMBIQUE

◄ *Most children in rural areas seldom attend school. Many work in the fields.*

◄ *More than four-fifths of the people, like these tea harvesters, work in agriculture.*

▼ *A woman of Mozambique Island hauls firewood and coconuts.*

➤ *Women outside Maputo tend to their small children while they tend their gardens.*

▼ *Once a vacation paradise for South Africans, Mozambique has few tourists today.*

➤ *Birds nest along the rivers and in the trees that border the grassy savannas.*

◄ *Many impoverished families live in primitive huts and own few possessions.*

▼ *Gorongosa National Park, in the central hills, is the nation's largest wildlife preserve.*

⋏ *Mozambicans celebrate Independence Day in June with parades and political rallies.*

➤ *Pelicans compete for food in the coastal waters.*

⋏ *A country school may be nothing more than an open-air courtyard.*

⋎ *Fishing boats crowd the entrance to a small harbor on Mozambique Island.*

recruited large armies of black troops to fight them there, using old tribal rivalries to turn the Mozambicans against each other. And Portuguese officers punished villages that aided Frelimo; for example, they ordered hundreds of people burned to death in the community of Wiriyamu in 1972.

By 1974, ten years after the beginning of the rebellion, Frelimo had won significant victories in central Mozambique. Dissatisfied with the war, with uprisings in two other African territories, and with their government at home, the officers of the Portuguese army staged a coup d'état. Frelimo took advantage of Portugal's turmoil to declare the People's Republic of Mozambique independent in June 1975. The group declared Machel president and changed the name of Lourenço Marques to Maputo, a traditional African word for "home village." Frelimo set up a Socialist government, which discouraged private ownership and gave the state control of all utilities, transportation systems, and medical and educational facilities.

The transition from colony to independent nation was not smooth. When independence was declared, all but 15,000 of the Portuguese administrators, equipment operators, and doctors left the country. Angered at the loss of what they considered to be their territory, many of the departing Portuguese destroyed buildings, machinery, food, and medicine. Only 40 doctors remained in the country, and only 7 percent of the population could read or write. The economy never fully recovered, and in 1984, facing debts totaling $1.6 billion, the country declared bankruptcy.

In addition to its serious social and economic problems, Mozambique faced a grave political situation. Its closest neighbors at the time of independence, South Africa and Rhodesia, were ruled by white governments that opposed African liberation. Both nations stopped using Mozambican laborers and transport services, and the country's income dropped sharply. When an independence movement began in Rhodesia, Frelimo supported it and provided a base for the nationalist guerrilla forces who were working to overthrow the white government. As a result, Rhodesian troops attacked Mozambique several times in the late 1970s, once mounting a full-scale invasion of Tete province.

African nationalist forces declared Rhodesia to be the independent nation of Zimbabwe in 1980, and today relations between Mozambique and Zimbabwe are good. The troubles with South Africa lasted longer. During his time in office, President Machel openly advocated sanctions (economic boycotts) against South Africa. In turn, South Africa accused Mozambique of harboring members of the African National Congress (ANC), the principal group dedicated to the overthrow of the South African government.

South Africa also supported the guerrilla war that the Mozambican National Resistance (Renamo) waged against Frelimo. That war began in 1976 and had a disastrous effect on Frelimo's plans for economic development. Renamo guerrillas carried out hit-and-run attacks on government installations, roads, bridges, and radio stations throughout Mozambique.

When South Africa made the transition to black majority rule in the first half of the 1990s, it stopped supporting Renamo. In 1992, Frelimo

Mondlane (right) and Machel (left) led the fight for independence from Portugal.

In 1975, Mozambicans in Maputo greeted independence by pouring into the streets.

and Renamo finally negotiated a cease-fire. The first free elections were held in 1994, and Frelimo retained control of the government. Joaquim Chissano of Frelimo received 53 percent of the vote for president. The leader of Renamo, Afonso Dhlakama, received 34 percent. Frelimo won 129 seats in the 250-seat parliament; Renamo won 112 seats. Hopes were raised that peaceful political competition would prevent the kindling of another civil war.

A food vendor in an outdoor market prepares cassava leaves for cooking. They will add flavor to a thick porridge made of cassava flour and corn.

Country of
the Good People

More than 96 percent of Mozambique's population is African. Another 1 percent is Swahili (people of mixed Arab and African ancestry). The remainder consists of Europeans (mostly Portuguese), Indians, Chinese, Pakistanis, and mestizos (people of mixed African and European ancestry). These non-Africans live in the coastal cities; they are usually doctors, teachers, shopkeepers, or industrial workers.

During the ten-year war of independence, many Mozambicans fled into Tanzania, Malawi, and Zambia. But by 1976, most of them had returned to Mozambique. According to the United Nations high commissioner for refugees, some 35,000 Mozambicans returned from Malawi and Zambia and more than 80,000 from Tanzania.

Mozambique's native African population draws from nine major ethnic groups. The largest group consists of the Makua-Lomwe people, who account for slightly more than half of the total population. They live in the northern part of the country, between Malawi and the Mozambique Channel. Most are subsistence farmers who cultivate crops for their own survival rather than to sell. A few, however, grow cash crops, and a significant minority are fishermen who have settled in villages along the coast.

North of the Makua-Lomwe live the Makonde and Yao peoples, whose relatives live in Tanzania. The Makonde, who live near the coast,

On the deck of their fishing boat, these Swahili fishermen sort the day's catch.

cling to ancient, traditional customs and ways of life. The Yao are the only inhabitants of the remote and thinly settled northwestern region near Lake Malawi. They were active in the slave trade during the 18th and 19th centuries and were greatly influenced by the Arabs. Most of them are Muslims.

All of the groups living north of the Zambezi River share certain cultural elements. They are matrilineal, tracing their origins and ancestry through their mothers' families rather than through their fathers'. Many of them move every few years to new, more fertile soil. Because they have always inhabited remote areas far from the capital city, they were less affected by Portuguese domination than were the southern Mozambicans.

The people who live south of the Zambezi River in the Zambezi Valley felt the effects of the Portuguese occupation more keenly because the Europeans were once concentrated in this region. As a result, the southern Mozambicans adopted European dress, languages, and religions more readily than did the northern Mozambicans.

✳ Most of the Mozambicans who worked in the South African mines came from the large Tsonga tribe, which lives between Delagoa Bay and the Save River. Other southern groups include the Karanga (descendants of the Mutapa), the Chopi (descendants of the "good people" who greeted da Gama on his arrival), the Shona, and the Nguni. Each of these groups traditionally consists of cattle-herders and farmers and is patrilineal, tracing ancestry through the father.

Nine-tenths of the country's population is rural, living on isolated homesteads and farms or in small villages. The villages are surrounded by *bomas*, fences made of tall poles with sharpened tips that prevent lions and other animals from entering at night. Traditional village homes are round huts with walls made of poles plastered with mud and roofs thatched with palm leaves. But small, square, tin-roofed houses made of cement blocks are becoming more common.

During the civil war many Frelimo soldiers lived in large tents, which were comfortable compared to the traditional small huts.

At the center of each village is either a kraal (cattle pen) or a community building, such as a school or agricultural training center. Outside the
village are the fields. According to African tradition, this land does not
belong to individuals but to the entire community. Since much of the
land is prone to drought and not particularly fertile, subsistence farming
can be a very labor-intensive activity.

The average population density of Mozambique is just 57 people per
square mile (22 per square kilometer). The most densely populated provinces are Nampula and Zambezia along the northern coast; they contain
40 percent of the country's population. Niassa province in the north is
almost empty as a result of slave raids in the 19th century, sleeping sick-
ness epidemics in the 20th century, and poor soil unfavorable to agriculture. The upper reaches of the Zambezi are also sparsely populated. The
new lake created by the Cabora Bassa Dam may one day improve the
fertility of this region and attract settlers from the more crowded coastal
provinces.

Thickly settled farmlands adjoin the coastal cities of Maputo, Xai-Xai,
Beira, and Angoche and often support large-scale production of fruit,
sugar, and rice rather than small, traditional farms. Many of these large
farms are former colonial estates or concessions that were nationalized
between 1975 and 1977. The government now maintains agricultural
training schools on them and hopes to interest more tribespeople in modern farming techniques.

When Mozambique became independent, Frelimo hoped to banish
the Portuguese language. However, it could not find an adequate replace-
ment, because no other language was spoken by a majority of Mozambicans. The people of the north speak the Bantu languages of Yao and
Makua. Those in the Zambezi Valley speak Nyanja, and many southerners
speak Tsonga. Mozambicans who live along the coast, especially in the
north, speak Swahili. So Portuguese has remained the official language; it
is taught in the schools and used to conduct government affairs, but the
people rarely speak it outside the cities. Many of those who do business

with their English-speaking neighbors in Zimbabwe and South Africa also speak English.

The constitution that Frelimo established in 1975 guaranteed all Mozambicans the right to practice any religion they chose. For a long while, however, Frelimo was visibly hostile to the Roman Catholic faith, regarding it as a remnant of the discarded colonial culture. Despite this era of disapproval, about 16 percent of the population has continued to worship in Catholic churches, especially in the south, and many missions that were founded during the colonial period still provide medical and educational help as well as religious instruction.

About 20 percent of Mozambicans, mostly in the north, are Muslims. Another 14 percent of the population consists of Protestants, who are scattered throughout the country. A small number of Hindus and Jews live in the coastal cities.

Built by the Portuguese in the 16th century, Misericordia Church on Mozambique Island is part of the country's Catholic heritage.

The remaining 50 percent of Mozambicans practice traditional tribal religions. Most of these tribal religions are animistic; they attribute good and evil spiritual qualities to things in the natural world, such as animals, trees, and weather, and teach that sorcerers, wise men or women, and witch doctors can communicate with these spirits. Some Mozambicans combine the dancing, chants, and rituals of their traditional religions with some form of Christianity. Many also combine the traditional herbal medicines of the village healers with the pills handed out by visiting government doctors.

During the colonial era, the administrators closed Mozambique's offices and businesses on Christian holidays, such as Easter and Christmas. Today's government does not recognize these religious occasions and has introduced a new set of national holidays instead. These include New Year's Day (January 1), Mozambican Heroes' Day (February 3), Mozambican Women's Day (April 7), Workers' Day (May 1), Independence Day (June 25), Armed Forces Day (September 25), and Family Day (December 25). All government offices and businesses are closed on these days. In addition, Mozambique continues to follow the colonial custom of the *siesta*, or midday rest period. Stores, schools, offices, and businesses close for three or four hours during the hottest part of the afternoon and reopen in the cooler evening hours. People usually sleep for several hours after lunch.

The food of Mozambique reflects the country's triple heritage: African, Arab, and Portuguese. Very few farmers today grow the grains cultivated by the original Africans, except for millet, which is sometimes made into beer. Instead, the chief food crops are corn and cassava, which the Portuguese discovered in the Americas and introduced to Mozambique, along with cashews, pineapples, and peanuts. All of these imports are now important crops, along with oranges and lemons from Arabia, ginger and mangoes from India, and rice from Indonesia.

Much of the rural Mozambicans' diet is based on the root of the cassava plant. In fact, the African name for cassava means "the all-sufficient." The nutritious root can be baked like a potato, ground into flour, dried in the sun like fruit, or mashed with water to form porridge. To make a typical dish for a country family, a woman first grinds corn and cassava together into coarse flour, using a wooden bowl and a heavy wooden pestle. Next, she adds to the mixture cassava leaves—similar to spinach—and water to make a yellowish dough. She serves the result in calabashes (hollowed gourds dried in the sun), accompanied by a handful of roasted nuts and washed down with weak palm wine, called *shema*.

A sharpened stake is used to split coconuts. The coconut meat seasons local dishes.

The coastal Mozambicans enjoy a greater variety of meals. Seafood is popular, and specialties include Portuguese-style dishes such as *macaza* (shellfish skewered on bamboo twigs and grilled over an open fire), *bacalhão* (dried, salted fish mixed with vegetables), and *chocos* (squid cooked in its own ink). The coastal people also eat more rice and fruit than those of the interior. Throughout the country, cooks use peppers, onions, and coconuts to season most foods.

Clothing is another aspect of Mozambican life that reflects influences from several cultures. In the cities and towns along the coast, men frequently wear Western-style suits. All government workers have adopted this type of clothing, usually in dark colors. In the streets and shops of the towns, the women wear dresses cut in Western styles but made of brightly colored and patterned African fabric.

Many Mozambican women, however, continue to wear traditional African dresses, especially in rural districts. These garments are long strips

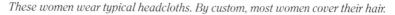

These women wear typical headcloths. By custom, most women cover their hair.

of material wrapped around the body under the arms and over one shoulder. Most also wear turbans, scarves, or headcloths.

Men throughout much of the country have abandoned the traditional loincloths in favor of shorts worn with Western-style T-shirts or *dashikis*, loose, pullover shirts of brightly printed cotton. Sunglasses are popular with men who can afford them. All but the poorest children now wear Western-style clothes, including many secondhand garments provided by missionaries or international aid societies. Most rural Mozambicans have only one or two sets of clothes.

Muslim men and women in the north wear long, white robes, turbans, and veils. Asian men in the coastal cities wear two-piece, white, cotton suits with high-collared jackets and Asian women wear black or colored silk dresses. The streets of almost any city or town are often filled with people of many races wearing clothes of different styles. Most Mozambicans still frown on certain Western styles however, such as blue jeans, long hair, and short dresses.

Culture and the Arts

The two most highly developed forms of traditional art in Mozambique are wood carving and dancing. Many Africans still practice both traditions, especially the Makonde people in the north. Makonde men transform hardwoods (chiefly ebony, mahogany, and ironwood) into statues and masks that are prized all over Africa.

Some of their designs reproduce the traditional tattoo patterns of the Makonde women. When they are teenagers, these women receive hundreds of tiny cuts on their faces, arms, and backs, arranged in zigzag geometric patterns. Repeated over the years and rubbed with charcoal, earth, and herbs, the cuts form raised scar tissue several shades darker than the skin. In the past, the Makonde regarded these patterns as the height of female beauty, but today the practice is beginning to die out. Some historians believe that the tattooing—as well as the Makonde custom of stretching the lips by inserting flat plates or plugs into them—arose

The Makonde men of the north are expert woodcarvers, who transform ebony and mahogany into prized masks and statues. Some of the masks are used in religious rituals.

centuries ago in an attempt to make girls and women less attractive to slave raiders. Government health officials now discourage tattoos and lip plugs, believing that they lead to infections.

The Makonde also have a long tradition of dance. Most of the dances are very old and have religious meanings. In one often-performed dance, men called *mapicos* wear huge wooden masks that they have carved and painted in secret; the masks represent demons, and women are forbidden to touch them. To the music of drums and antelope horns, a series of mapicos pretend to attack the clustered villagers. Each time the villagers chase the demon away, another takes his place. The complete dance lasts for many hours.

The Makua people of north central Mozambique perform a dance on stilts. Wearing comical masks and dressed in their most colorful clothes, men strap 6-foot (2-meter) poles to their legs and drape the stilts in lengths of red and yellow cloth. With great skill and endurance, the men jump, dance, and strut around the village on the stilts—sometimes for hours. The women of Mozambique Island practice another form of dancing, combining dance steps with rope jumping to a fast drumbeat.

The Chopi people of the south preserve a traditional hunting dance. Dressed in lion skins ornamented with monkey tails, warriors carrying spears and large, oval shields circle around fires in mock attacks and counterattacks. The Chopi accompany their songs and chants with the music of *marimbas* (hollow gourds filled with nuts or pebbles that rattle when shaken) and *mbiras* (strips of scrap metal mounted on a hollow drum or box and plucked with the fingers). Some of their songs are similar to West Indian calypso or reggae music. Mozambicans of the central and southern regions compose many songs, which often make fun of their neighbors or current events.

The Frelimo government has established the National Institute of Culture to collect and record native music, crafts, stories, and mythology. The institute maintains cultural centers in the urban communities and the larger villages.

✳A few Mozambican artists have become well known in other parts of the world. One is painter Malangatana Goenha Valente, who crowds his large canvases with figures and images that represent the struggle between African beliefs and European culture. One of his paintings, for example, shows a witch doctor and a priest battling over the soul of a dead African. Valente's work hangs in many museums around the world.

✳Mozambique's poverty and low literacy rate limit communications.

Collective Work, *a painting
by Agostinho Mulemba,
shows workers on a
communal farm.*

The country has only two newspapers, which about 81,000 people read daily. One out of every 21 people in the country owns a radio, 1 in every 251 owns a telephone, and 1 in every 281 owns a television.

*Samora Machel headed the Frelimo party and Mozambique from the granting of
independence until his death in 1986.*

Government and Social Services

Mozambique's ten provinces are divided into about 94 districts and many smaller townships, called *localidades*. All adult men and women vote to elect representatives to assemblies in their localidades. These representatives, in turn, elect representatives to district assemblies.

When Mozambique became an independent state in 1975, Frelimo turned it into a one-party socialist state. But in the late 1980s, the government began to introduce economic and political changes. In 1990 the country received a new constitution that provided for a multiparty political system, a market-based economy, and free elections. The first elections were held in 1994, two years after Frelimo signed a peace agreement with Renamo.

Under the 1990 constitution, Mozambique is a republic with an elected president and an elected legislative branch, the Assembly of the Republic. The president and the members of the Assembly serve five-year terms. The president is the chief of state. The head of the government is the prime minister, who is appointed by the president. The government also includes a cabinet, similar to the United States cabinet. The cabinet ministers run departments such as finance and defense.

The police force is a national organization, with branches in every city and town. Mozambique's armed forces are called the FADM—for Forces Armadas de Defesa de Mozambique. The FADM was formed at the end of the civil war with Renamo. Under the terms of the peace agreement, the old Mozambican military forces were disbanded, and troops from both sides joined the FADM. The new force was supposed to number 30,000 troops, but only about 11,000 troops from the old armies decided to enlist. Today, the FADM has about 15,000 troops.

Health Care

Health services collapsed at the end of the colonial period, when nearly all skilled doctors and nurses fled the country. Frelimo immediately set up a

After Machel was killed in an airplane crash, Joaquim Chissano was named head of Frelimo and president of the country.

crash program to train paramedics—medical assistants capable of providing routine health care and emergency first aid. With technical and financial help from the United Nations Children's Fund and the United Nations Development Program, the northern provinces of Cabo Delgado, Niassa, and Tete carried out vaccinations against measles and smallpox.

Mozambique has only 435 doctors (1 for every 41,000 people) and 13,320 hospital beds (1 for every 1,300 people). Measles, tuberculosis, hepatitis, leprosy, cholera, and tetanus are the chief causes of death. Many Mozambicans also suffer from malaria, the tropical disease that is carried by anopheles mosquitoes. The average lifespan of Mozambican men is 44 years; the lifespan of women is 45. The infant mortality rate is one of the highest in the world; for every 1,000 babies born, 125 die during infancy.

The end of the civil war opened up greater opportunities for health care progress. The government has developed a new health care plan in cooperation with the World Health Organization (WHO) and the World Bank. Over the next 10 to 20 years, spending on public health is supposed to increase 4.5 percent per year. Improvements in rural health services have been given a high priority. The staffs of rural district hospitals are receiving special training in basic surgery and anesthetics. Finland is funding a 12-year, $20-million program to upgrade health care in Manica Province. Five years after the war ended, the World Health Organization reported that the effort to improve rural hospitals was already showing results. In particular, more mothers were surviving childbirth in the hospitals where the staff had received the extra training.

Besides hospital services, preventive health care is a crucial consideration—that is, helping people to avoid illness rather than waiting until they are sick to cure them. Even before the latest programs began, Frelimo arranged for teams of trained volunteers to go into rural villages not only to vaccinate the children but also to teach adults the fundamentals of sanitation, nutrition, and first aid. A great deal of work remains to be done, however, before adequate health care is available throughout the country.

Education

The law in Mozambique requires children to attend school for seven years, but only about 41 percent of the children do so. Only 7 out of every 100 children go on to secondary school. In rural districts, children, especially girls, seldom attend school. One of the government's goals is to promote education for girls and women. Like health care, however, education in Mozambique has been hampered by shortages of materials and trained people.

In addition, the fighting between Frelimo and Renamo often disrupted schooling in some parts of the country. In 1979, when the civil war was still in its early stages, 52 percent of the school-age children in Mozambique were enrolled in primary and secondary schools. By the time

The government hopes to improve Mozambique's low literacy rate.

the war ended, enrollment had dropped to 32 percent. The nation is still recovering from that disastrous period.

Today, the country has approximately 3,770 primary schools, 240 secondary schools, 32 technical or vocational institutions (where students learn agricultural and mechanical skills), and only 1 university: Eduardo Mondlane University in the capital city, Maputo. More than 5,000 students are now enrolled in the university.

Agriculture makes up a third of the country's yearly economic production, but droughts and poor soil often cause meager harvests.

Economy
and Transportation

During the colonial period, Portugal wanted Mozambique to serve as a market for inexpensive goods manufactured in Portugal. As a result, it did not encourage industry and manufacturing in the colony, even though Mozambique had some significant mineral and energy resources. The bulk of Mozambique's foreign earnings came from labor exports and transportation services, and the country's internal economy relied on agriculture. Since independence, Mozambique's economy has suffered greatly under the combined forces of civil war, drought, and famine. In 1995 alone, the nation received over 400,000 tons of international food aid.

Agriculture still accounts for about 33 percent of the country's total economic production every year and occupies about 90 percent of the work force. Unfortunately, Mozambique is not a very fertile country. Water shortages and poor soil often lead to poor harvests; a particularly bad drought that lasted from 1981 to 1984 caused many crops to fail. The most fertile areas are the lowland river basins and the *machongas*, patches of rich soil between the sand dunes of the coastal plain.

Farmers grow cassava, peanuts, corn, vegetables, and sorghum for use at home and cashew nuts (Mozambique is the world's largest producer of these), tea, copra (coconut meat, used in vegetable oils, soaps, and cosmetic products), sugarcane, bananas, and citrus fruit for export. Log-

Roundwood logged in the forests of Manica Province is sold to builders in South Africa.

gers harvest roundwood in the forest areas and sell it to South Africa as timber.

The country's farmers have about 1.25 million cattle and 22 million chickens. Every few years, tsetse flies infect the cattle herds with epidemics of a disease called *rinderpest*. As a result, most of the cattle are thin and unhealthy. Many tribespeople, especially in the south, consider cattle to be a form of wealth and do not often eat their beef.

Other Mozambican products include large quantities of shrimp from the coastal waters and some cotton produced on the old colonial plantations. For its exports, the country's biggest customers are Spain, South Africa, the United States, Portugal, and Japan. South Africa's purchases include some of the electrical power generated by the Cabora Bassa hydroelectric plant.

Mozambique purchases large quantities of food, clothing, farm equipment, and petroleum from South Africa, the United Kingdom, Japan, and Portugal. Each year the nation imports more goods than it exports. To offset this unfavorable balance, Mozambique still depends heavily on aid from other countries in the form of cash grants, loans, food, and medical supplies. The nation's total foreign debt remains high.

Overall, by the mid-1990s, some signs of hope were beginning to appear in the official trade figures. Experts realize, however, that the official figures tend to underestimate Mozambique's total foreign trade. The official figures do not include a very active "informal sector" that conducts most of the trade that passes between Mozambique and the countries on its borders.

During its socialist period, Frelimo built state-owned factories to produce bicycles, soft drinks, textiles, clothing, soap, and cement. Its larger industrial projects included a steel mill at Beira and an oil refinery, a chemical plant, and a fertilizer factor at Maputo. Many of these enterprises have been privatized (transferred to private companies) or will be privatized in the future. Some of the privatized businesses have been successful; others have failed, sometimes because of lack of good equipment.

Frelimo has hired foreign experts to continue geological research begun by Portuguese scientists before 1975. Surveys of the western part of the country show that it has the potential for a prosperous mineral industry: the experts have located deposits of more than 700 million tons (777 million metric tons) of coal, as much as 360 million tons (400 million MT) of iron ore, and small amounts of manganese, asbestos, gold, diamond, graphite, mica, and uranium. Mozambique is also believed to have the world's largest supply of tantalite, a rare and valuable metal used in certain kinds of computer and space equipment.

In terms of energy resources, Mozambique possesses sizable reserves of petroleum and natural gas, as well as many sites along rivers that would be good locations for hydroelectric power plants. If it can tap these resources, Mozambique will have enough energy both to meet its needs and to sell power to the neighboring countries.

Tourism is another resource the country hopes to exploit in the coming years. In the 1950s, certain spots along the coastline became popular as vacation resorts for whites from Rhodesia and South Africa. Chief among them were Benguerua and Bazaruto, two small coral islets near Mozambique Island, where visitors enjoyed scuba diving and fishing for marlin. The shops and restaurants of Lourenço Marques and the wildlife safaris of Gorongosa National Park also brought tourist income into the area. During the war for independence, however, the small tourist industry collapsed. Few travelers visited Mozambique during the guerilla fighting. With the end of the civil war, the potential for Mozambique's tourist industry soon began to attract foreign investors.

Mozambique's unit of currency is the *metical*. A metical, which is made up of 100 centavos, is worth about a hundredth of a U.S. penny; about 11,000 meticais (pronounced met-a-CASH) equal U.S. $1. Prices for food and clothing are high, especially in the cities. For example, a meal of fish and potatoes at a restaurant or outdoor food stall in a coastal city costs about 52,000 meticais, which is the amount the average Mozambican worker earns for four or five days' work. Urban workers earn more than the average, and rural workers earn less—although most of them grow their own food. Food shortages are fairly frequent in the cities and towns.

Transportation

The country's transportation system serves primarily to ship goods to and from neighboring nations, rather than to carry passenger and freight traffic within Mozambique. Railways are the most important feature of the system. Mozambique's rails link other African nations to the sea. One network of rails connects Maputo with the Witwatersrand in South Africa and the Limpopo Valley. Another connects Beira with Zimbabwe, Malawi, and Zambia. A third runs between Nacala and Malawi. These three systems—together with smaller rail lines that serve the coastal communities—run on a total of 1,945 miles (3,130 kilometers) of track and carry about 336 million ton-miles (616 million metric ton-kilometers) of cargo each year.

Mozambique has about 16,950 miles (27,280 kilometers) of roads. The main coastal highway from Pemba to Maputo is paved, as are three highways linking the coast with the bordering countries. They are used chiefly by cargo trucks, although the country does have 27,000 passenger cars. The unpaved, secondary roads are rutted and bumpy during the dry season, and swampy and muddy during the wet season. Many sections of the

Cabora Ba~ ... ~umwezi River, is one of Africa's largest hydroelectric plants.

country have no real roads, only tracks for foot travelers or cyclists (bicycles are a popular form of transportation).

Maputo, Beira, and Nacala are the chief seaports. The government has installed special equipment for loading iron ore and coal into cargo ships in the harbor at Maputo, which is large enough to hold 20 oceangoing vessels at once. Beira's harbor is shallower and is used mainly by fishing craft and smaller cargo vessels, such as coastal steamers. The harbor at Nacala is large and modern; the government expects it to grow if mining and industry increase in the north. Smaller ports are located at Pemba, Mozambique Island, Quelimane, Inhambane, and Xai-Xai.

Government, medical, and business workers who have to travel within the country do so by air. Every town and city has a small airport, and the bush country is dotted with tiny landing strips.

This boy selling coconut candy is one of Mozambique Island's many vendors. The island combines age-old charm with busy trading and fishing industries.

Cities and Towns

Mozambique's largest cities are Maputo, Nampula, and Beira. All three are growing rapidly, as are most other urban centers in the country. Rural Mozambican men are drifting to the towns and cities in search of jobs; girls and women, however, tend to remain on the farms. Thus, families are separated, and the crime rate of the cities increases when the uneducated newcomers are unable to find work. To combat the problem, Frelimo is trying to discourage young men from migrating to the cities until they have completed their educations.

Most of the people who come to the coastal cities from the country settle in shantytowns—crowded ghettos in the suburbs. Poor sanitation there leads to disease that can quickly spread throughout entire provinces. Government planners hope to replace the shantytowns with low-cost apartment buildings, but the supply of apartments cannot keep up with the growth of the cities.

Maputo, the former colonial capital of Lourenço Marques, has a population of 1,700,000. It is a gracious and attractive city. It was planned and built on the European model, with wide boulevards, public gardens and parks, paved sidewalks decorated with mosaic tiles in wavy patterns, and impressive, marble-fronted public buildings. Maputo is actually two cities in one: the houses, stores, and high-rise apartment buildings stand on a

The capital's name, Maputo, is a traditional African word that means "home village."

steep bluff overlooking the harbor, while the port facilities, factories, and most office buildings are concentrated between the foot of the bluff and the ocean.

Between the two halves of the city is an open ravine where coal is stockpiled before it is shipped. On the outskirts of the city is a semicircle of small, private houses. Most of them are two- or three-story villas painted white, with red-tile roofs imported from Portugal. They were once the homes of Portuguese administrators and businesspeople but now house Mozambican shopkeepers and workers. Beyond them is the shantytown.

The architect of many of the most striking office and apartment buildings in Maputo was Amancio d'Alpoim Guedes, a Portuguese who received worldwide attention for his work in Mozambique during the 1950s. Inspired by traditional African art, Guedes tried to add shapes and symbols from native architecture to his modern constructions. Long spines protrude like poles from the outer corners of Guedes's buildings, chimneys are shaped like mushrooms, and wall mosaics feature African designs.

With a population of 300,000, Beira is the second largest city in the country. Like Maputo, it was built by the Portuguese in the spacious

European style. Beira is the focus of trade with Malawi and Zimbabwe, as well as the center of Mozambique's commercial fishing industry.

Nampula, Mozambique's third largest city, has 230,000 inhabitants although it is only half a century old. The Portuguese drained a swamp and built schools, stadiums, churches, offices, and houses, hoping that Nampula would become the metropolis of northern Mozambique. It has grown even faster than the colonial administrators expected. Nampula is located midway along the commercial railway between Malawi and the port of Nacala. Farming communities in the Lake Malawi area ship their goods to market from Nampula, and many banks, trading companies, and businesses have offices there.

Angoche (called António Enes in colonial days) is a typical, small, coastal town, with a population of several thousand. It lies at the foot of a small hill that, although only 200 feet (66 meters) tall, is the highest point for a long distance along the coast in either direction. Rows of small, boxlike houses, all painted white to reflect the hot tropical sun, border Angoche's tree-lined streets. The people of the town are fishermen, who use 30-foot (10-m) canoes, called *almandias*, made from giant logs and fitted with small sails. When evening comes, the fishermen sail in from the ocean and pull their almandias up onto the coral sand for the night. In the adjoining townships of Inguri and Puli, about 20,000 people live in traditional, African-style villages and practice subsistence agriculture. The villagers trade vegetables to the Angocheans in return for fresh fish.

Mozambique Island is the oldest surviving settlement in the country. Coral-block mosques and houses built by the Arabs in the 12th and 13th centuries still stand. So does Fort St. Sebastian, an immense stone fort that the Portuguese built when they took over the island in 1507. Armed with cannons, the fort withstood attacks by the Arabs, Dutch, and French. The Portuguese named each of its towers for a Catholic saint and built a tiny chapel just outside its walls called Our Lady of the Bastion. Legend has it that the deep cisterns in the fort's basement could hold enough water for 1,000 men for a full year. Today, Mozambique Island preserves

The market is an important part of every Mozambican community, large or small.

Fort St. Sebastian as a museum. The town itself is small but busy. Many of the country's Asian and Muslim inhabitants live there and work at trading in the open-air markets and shipping in small, coastal craft on the Indian Ocean.

Some of Mozambique's most interesting cities are completely uninhabited. One such is Nhacangara, an ancient mountaintop fortress near the Zimbabwean border. Shaded by thick trees and partially covered by vines and undergrowth, Nhacangara consists of massive walls of fieldstone and many narrow paths and tunnels around the peak. Stone pillars outside the crumbling walls may have been supports for long-vanished wooden roofs, or they may have served some other purpose. Traces of terraces once used for gardens and farm plots remain on the sides of the nearby hills. Bits of clay pottery, their painted designs almost obliterated, lie among the fallen stones.

Bats, snakes, and owls are Nhacangara's only residents now, but centuries ago it was an outpost of the Mutapa empire. The captains of the fort gathered gold from the miners and sent it through the jungle in slave caravans, to be traded to the Arabs of the coast. Archaeologists speculate that many more such forts and cities remain to be discovered and studied in Mozambique's wilder regions.

Despite tattered clothes and homemade toys, these boys demonstrate the Mozambican spirit of friendly good humor that da Gama's men appreciated 500 years ago.

Mozambique
in Review

Located on Africa's southeastern coast, Mozambique has a population of about 18,000,000—roughly the same as the state of Texas. Most of the country's inhabitants are members of ten African tribal groups, but the population also includes many people of Arabic, Portuguese, Chinese, and Indian descent. Modern Mozambican culture combines traditional African, Arabic, and Portuguese elements.

Throughout its history, Mozambique has been a crossroads for travelers and traders from the African interior, Arabia, Europe, India, and Southeast Asia. For centuries, powerful African empires flourished in the areas that later became the nations of Zimbabwe and Mozambique. Although these empires left no written history, archaeologists have begun to discover some of their secrets by exploring the ruins of ancient stone cities. They now know that Arab traders reached Mozambique as early as the 8th century A.D. and that, by the 11th century, they had established trade routes linking their cities along the African coast to the gold mines and cities of the interior. These traders controlled a network of sea and land traffic that carried goods and slaves between Africa, India, and the Far East.

The first Europeans to explore Mozambique arrived in 1498, with Portuguese navigator Vasco da Gama. In the 16th century, the Portuguese drove out the Arabs and took control of trade in Mozambique. They began trying to conquer the local African kingdoms as early as 1569, but the native people—especially those who lived in remote highlands in the heart of the country—fought back fiercely. Portugal easily took control of the coastal areas but had a hard time subduing the Mozambicans of the interior. Tribal rebellions against Portuguese rule continued until 1917, when the Makonde people participated in the last native uprising.

In the 1960s, a group of nationalists who believed Mozambique should be independent organized the revolutionary Mozambique Liberation Front, or Frelimo. After more than ten years of fighting, Frelimo forced Portugal to give up its hold on Mozambique in 1975. The Portuguese administrators withdrew and turned the colony over to the Mozambicans, who named their new country the People's Republic of Mozambique.

After winning the fight for independence, Frelimo promptly banned most foreign investment because it wanted Mozambicans to receive all of the profits from economic activities in their own country. But since 1975,

Mozambique's leaders are fighting poverty and hunger to build a stable future.

Frelimo has recognized that its original policies were a failure and that Mozambique needs outside financing if its economy is to prosper. The country has been steadily moving toward a free-market economy since the late 1980s. The process has accelerated since a new constitution was enacted in 1990. By the late 1990s, over two-thirds of Mozambique's industrial output was being produced in the private sector.

In 1976, one year after Mozambique became independent, Frelimo became involved in a civil war with a rebel organization named Renamo. In the beginning of the war, Renamo was supported by the white governments of neighboring Rhodesia and South Africa. In 1980, a black government took control of Rhodesia, renamed the country Zimbabwe, and withdrew support for Renamo. A similar process occurred in South Africa, but not until the 1990s. The civil war, therefore, dragged on from the late 1970s into the 1990s. Its effects on Mozambique's economy and society were devastating.

Finally, in 1992, Frelimo signed a cease-fire with Renamo, and the civil war finally came to an end, 16 years after it had begun. In 1994, Mozambique held its first free elections and installed its first elected government.

Today, Vasca da Gama's Country of the Good People is a land that faces many difficult tasks. It must increase its food production. It must develop its economy and improve its educational system. But the people of Mozambique can now look to the future with some confidence. For the first time since gaining their independence from Portugal, they can concentrate on their problems without being harassed by civil war and hostile neighbors. Proud of their independence, Mozambicans hope to see their country develop into a successful modern nation.

◄GLOSSARY►

Almandias	Fishing canoes made from hollowed logs fitted with sails.
Animism	The philosophy, held by many African peoples, that all natural things—people, plants, animals, and stones—possess souls, through which the gods watch and influence everyday human life.
Assimilado	Under the class laws of 1927 to 1961, a native African who shared some of the civil rights of white citizens by giving up his African lifestyle, speaking Portuguese, and taking a job in a factory or business.
Bacalhão	A Portuguese-style dish of dried, salted fish and vegetables.
Baixos	Floating islands of vegetation that endanger water traffic in the Zambezi Delta.
Boma	A fence of sharpened wooden stakes around a village or homestead.
Chikunda	The private army of a prazero, or landholder, whose soldiers were Portuguese and African.
Chocos	A dish made of squid cooked in its own ink.
Copra	Dried coconut meat used in foods and cosmetics.
Dashiki	A man's loose, pullover shirt of colored cotton.
Dhow	The traditional trade and fishing boat of the Indian Ocean Arabs, which has a distinctive, triangular sail.
Frelimo	The shortened name for the Mozambique Liberation Front, the country's ruling political party.

Grupos dinamizadores	Literally "dynamizing groups" in Portuguese. These societies promote Socialist theories and Frelimo policies in villages.
Indigenas	Most native Africans, as defined by the class laws conceived between 1927 and 1961. Indigenas were subject to special rules and had limited rights.
Kraal	A cattle pen, usually located in the center of a village.
Localidades	Small subdivisions of the 94 districts, similar to townships in the United States.
Macaza	Shellfish skewered on bamboo twigs and roasted over an open fire.
Machongas	Small, fertile patches of rich soil among the sand dunes of the coastal region.
Mapico	A male performer who enacts ceremonial dances while wearing a mask.
Marimba	A rhythm instrument made of a hollow gourd filled with pebbles.
Mbira	An instrument made of scrap metal mounted on a hollow box or drum and plucked with the fingers.
Mopani	A hardwood tree that produces durable timber.
Mwene	The title of the Mutapa king.
Não indigenas	Europeans, Asians, and African assimilados who enjoyed full civil rights under the class laws enacted between 1927 and 1961.
Prazero	A powerful landholder who owned a vast estate and commanded a private army sometime in the 17th through the 19th centuries.
Prazo	The estate of a prazero.
Renamo	Shortened name for the Mozambican National Resistance, once a guerrilla group, now a political party.
Savanna	A tropical or subtropical grassland containing scattered trees and drought-resistant undergrowth. Savan-

nas, also called "veld" or "veldt," are usually home to herds of grazing animals.

Senegal *khaya* A large tree that yields a dark, hard wood similar to mahogany.

Shibalo The system of forced labor by which the Portuguese ran their large colonial plantations, whose name derives from the Swahili word *shiba* (serf).

Swahili The ethnic group, language, and culture formed by the intermarriage and economic ties between Arabs and native Africans along the Indian Ocean coast.

◄ I N D E X ►

A

agriculture 61, 64, 66, 81– 82, 89
almandias 89, 97
animal life 26–29
animism 66, 97
Arabs 33, 34, 35, 38, 41
area 19
arts 69–73
assimilado 47, 97
average life expectancy 77

B

bacalhão 22, 68, 97
baixos 97
Bantu tribes 32–33
baobab 25
boma 63, 97

C

Cabora Bassa Falls 22
Cabral, Pedro Álvares 35
Cape Delgado (Cabo Delgado) 22, 33, 77
cassava 66–67
chikunda 37, 39, 97
Chissano, Joaquim 13, 16
chocos 68, 97
cities 87–91
climate 24–25
clothing 68–69
colonialism 43–57, 77, 81

communications 72–73
copra 97
cuisine 66–68
currency 10, 84
cyclones 24–25

D

dance 71
dashiki 69, 97
Delagoa Bay 22, 35
dhow 97
Dhlakama, Afonso 59

E

economy 81–84
education 78–79
energy resources 82, 83
England (see Great Britain)
ethnic groups 61–63

F

Frelimo (Mozambique Liberation Front) 13, 15, 16, 17, 48, 57, 59, 71, 75, 76, 77, 78, 87, 94, 95, 97

G

Gama, Vasco da 11, 15, 22, 33–35, 63, 93, 95
geography 9, 19–24
government 75–77
Great Britain 41, 44
grupos dinamizadores 76, 98

H
health care 76–77
holidays 66
hunter–gatherers 31–32

I
independence 13, 15, 17, 46, 47, 57,
   64, 94, 95
Indian Ocean 19, 24, 38
indigenas 46, 98
industry 82–83

K
kraal 64, 98

L
Lake Malawi 19, 20, 26, 33, 38, 40, 41,
   62, 89
Lake Oliveira Salazar 20
language 9, 33, 64–65
Lembobo Mountains 20
Limpopo River 21, 28, 34
literacy rate 9, 72
Livingstone, David 39–41
localidades 75, 98

M
macaza 68, 98
Machel, Samora 13, 15, 16, 48, 57, 58
machongas 81, 98
Madagascar 19, 23, 38, 43
Malawi 20, 44, 61, 84, 89
mapico 71, 98
Maputo 22, 24, 57, 84, 85, 87, 88
marimba 71, 98
mbira 71, 98
mestizos 61
minerals 83
missionaries 39–41
Mondlane, Eduardo C. 13, 48

mopani 25, 98
Mount Binga 20
Mozambican National Resistance (see
   Renamo)
Mozambique Channel 19, 22, 23, 61
Mozambique Conventions 45–46
Mozambique Island 22, 33, 34, 35, 43,
   45, 71, 83, 89
Mozambique Liberation Front (see
   Frelimo)
Mutapa (Matapa) 33, 35, 63, 91
mwene 33, 35–36, 98

N
não indigenas 47, 98

P
plant life 25–26
population 9, 61, 63, 87, 93
   density 9, 64
Portugal 15, 35, 41, 43, 44, 45, 46, 48,
   81, 82, 88, 94
Portuguese 34–39, 41, 45, 57, 62, 88,
   89, 93, 94
prazero 37, 43–44, 98
prazo 37, 98

R
religion 9, 33, 39, 62, 65–66
Renamo (Mozambican National Resistance) 13, 16, 17, 58, 59, 75, 76, 78,
   95, 98
Rhodesia (see Zimbabwe)
rinderpest 82
rivers 21

S
savanna 25, 98
Save River 21, 26, 63
Senegal *khaya* 25, 99

Serra da Gorongosa 20, 28
shantytowns 87, 88
sheiks 33, 34
shibalo 45, 99
Shire River 21, 40
siesta 66
slave trade 37–39, 41, 62
South Africa 19, 20, 28, 31, 39, 45–46, 57, 58, 59, 62, 82, 83, 84, 95
Soviet Union 16, 17
Swahili 33, 34, 35, 61, 99
Swaziland 19, 20

T
Tanzania 19, 48, 61
tattooing 69–70
Terra da Boa Gente ("Country of the Good People") 15, 61, 63, 95
tilapia 27
tourism 83

trade 33, 82
transportation 84–85

U
United Nations 16, 61
United States 16, 75, 82

V
villages 63–64, 71, 89

W
World War I 46
World War II 48

Z
Zambezi River 19, 21, 22, 28, 33, 35, 37, 39, 45, 62, 64
Zambia 19, 39, 40, 41, 61, 84
Zimbabwe 19, 20, 31, 33, 44, 46, 57–58, 83, 84, 90, 93, 95
Zanzibar 38

ACKNOWLEDGMENTS

The author and publisher are grateful to the following sources for photographs: Brian Seed/Click Chicago (pp. 2, 18, 23, 36, 50–51, 56b, 65, 67, 68, 86); FPG (pp. 52, 90); Library of Congress (pp. 25, 30, 32, 34, 38, 40, 44, 47); Mozambique Information Office (pp. 21, 58, 59, 60, 62, 70, 72–73, 76, 82, 85, 92); Ray Ellis/Photo Researchers, Inc. (pp. 49, 55a); Smithsonian Institution/National Museum of African Art (pp. 27, 29, 54b, 55b, 70); United Nations (pp. 17, 52–53, 56a, 74, 78, 88); World Vision (pp. 14, 42, 50, 54a, 63, 80, 94). Photo Editor: Mary Baldessari. Photo Research: Alan Gottlieb.